This book was written on the traditional
and unceded land of the
Anishinaabe Algonquin first nation.

It was illustrated and edited on
Treaty 1842 land,
the traditional territory
of the Baxoje Nation.

ISBN: 978-1-7352026-2-4

Contact us at:

danthefishbooks@gmail.com

Dana Digs Dinosaurs

Book One: Meet Dana

Written by Dana Korneisel
Illustrated by Carmen Cerra

Hi! My name is Dana,
and I love questions!

When I was little, I asked my parents and teachers lots of questions.

I especially liked to ask them questions about dinosaurs, but they didn't always know the answers.

I would sit on the scratchy carpet
of my school's library,
going through all the books about
fossils and dinosaurs.

I was looking for answers
about what I wanted to know.
Every time I got a new book,
I would find some more information
and think of more questions.

What happened to the extinct animals?

Are these pictures how they really looked?

How do scientists know animals are all related?

When I wasn't studying dinosaurs,
I loved to do a lot of other fun things.

I liked to wade in the creek and catch snakes and frogs, ride my bike with my sister and brothers, and read late at night under the covers.

Now that I'm grown up,
I still have fun doing new things.
I like to explore new places,
do circus arts and yoga,
and learn new crafts.

But my favorite thing to do
is study fossils.

I research giant fish fossils
from ancient swamps . . .

Tiny salamanders and fossils of our shared ancestors . . .

And even fluffy-feathered dinosaur fossils to learn about how fossils are formed.

I am a paleontologist!

Being a paleontologist means that I study fossils to understand the history of the Earth and answer questions like the ones I asked as a kid.

When I was little, I loved dinosaurs more than anything in the entire world.

But I never knew how many other kinds of fossils there could be!

Fossils aren't just dinosaur bones.

Fossils can be the bones and teeth of lots of other animals, too!

Fossils can be the shells of animals with no bones.

Fossils can be a space in the rock in the shape of an extinct animal whose body is gone.

Fossils can be footprints in a rock
which was once soft mud or sand.

Fossils can be minerals on a rock that show us what the soft parts of living things looked like.

Fossils can show us skin, feathers, beaks, leaves, and even hair!

With science, we can look at this kind of fossil and figure out some of the colors of extinct animals.

Some dinosaurs had red and black feathers.

Some laid blue eggs, or eggs with speckles on them, just like birds- the only dinosaurs still alive today.

Studying fossils helps me answer the questions I have and think of new ones.

My other scientist friends think of questions, too, about things I would never imagine.

That's one thing I love about being a paleontologist!

Definitions:

Paleontologist – a scientist who studies fossils

Research – the steps a scientist takes to gather new information and learn possible answers to our questions

Fossils – parts of a living thing or evidence of its behavior (footprints, burrows) left in a rock

Extinct – living things which lived in the past but are not around today

Questions I have for you!

What questions do you have about fossils?

What kind of research can you do to answer them?

If you become a scientist when you grow up, what would you want to study?

Dana answers your questions:

How are birds dinosaurs?

Scientists who study living things group them based on how they are related. A proper scientific group should include an ancestor and all of its descendants without taking any descendants out of the group. Think of it like a family tree. Your family includes your parents, you, and all of your siblings. If you made a group that excluded one of your siblings, it wouldn't be your whole family. The same is true for dinosaurs. Birds are the descendants of one group of dinosaurs, so to define the whole dinosaur family, we include birds.

How can you find color in a fossil?

Some colors are made of pigments. Pigments are specific materials found in a cell. The cells that make up your body and the bodies of many other living things contain miniature parts called organelles. They work in the cell similarly to how your organs work in your body. There are some organelles that you might know about like the nucleus and mitochondria. Others, you might not know. There are organelles called melanosomes that have color pigments in them. Melanosomes with different colors in them have different shapes. Sometimes, melanosomes are visible under a microscope on a fossil. We can look at their shape and compare them to the melanosome shapes in living animals to find out what color extinct animals were! Some colors we can identify in fossils are black and reddish-brown.

About the Author

Dana Korneisel is a PhD candidate at Carleton University in Ottawa, Canada. She is from Michigan and went to college at Iowa State University. After getting her bachelors degree, Dana went to Virginia Tech for her masters – where she studied the preservation of fossils thought to contain fossil blood cells. The results of this research are open access, meaning anyone can read the paper for free, in the journal PeerJ where they are published.

Now, she studies the evolution of the neck, especially in the first vertebrates to move from water to land, called tetrapods. In addition to being a paleontologist, Dana is also an enthusiastic aerial acrobat, reader, crafter, and is sure to be even more things in the future as she finds new hobbies. She would love to answer any questions you have about ancient life and what it is like to be a paleontologist!

About the Illustrator

Carmen Cerra earned his degree in Biological Premedical Illustration at Iowa State University in 1997.

He spent 15 years as a political cartoonist and copy chief before setting out on his own as a freelance illustrator.

He has done illustrations for wargames, tabletop roleplaying games, album covers, and personal commissions.

Dana Digs Dinosaurs is the first children's book he has illustrated.

Made in the USA
Monee, IL
05 November 2026

67380371R00024